BUILDING BLOCKS OF BIOLOGY

BACTERIA, FUNGI, AND VIRUSES

Written by William D. Adams

Illustrated by Ruth Bennett

WORLD BOOK

www.worldbook.com

Co-published by agreement between Shi Tu Hui and World Book, Inc.

Shi Tu Hui
Room 1807, Block 1,
#3 West Dawang Road
Chaoyang District, Beijing 100025
P.R. China

World Book, Inc.
180 North LaSalle Street
Suite 900
Chicago, Illinois 60601
USA

Library of Congress Control Number: 2025942738

Building Blocks of Biology
ISBN: 978-0-7166-6737-7 (set, hard cover)

Bacteria, Fungi, and Viruses
ISBN: 978-0-7166-6743-8 (hard cover)

Also available as:
ISBN: 978-0-7166-6763-6 (e-book)
ISBN: 978-0-7166-6753-7 (soft cover)

WORLD BOOK STAFF

Editorial

Vice President
Tom Evans

Senior Manager, New Content
Jeff De La Rosa

Proofreader
Nathalie Strassheim

Graphics and Design

Senior Visual Communications Designer
Melanie Bender

Acknowledgments
Writer: William D. Adams
Illustrator: Ruth Bennett/The Bright Agency

TABLE OF CONTENTS

There is a glossary on page 40. Terms defined in the glossary are in type **that looks like this** on their first appearance.

A MYSTERIOUS INFECTION
I'm sorry to hear that Root is sick!
Maybe we can cheer her up...
What's wrong, Root?
GET WELL SOON!
1
It feels like an infection, but I don't know what kind of germ it is...

Root looks awful. I wish there was something we could do...
ROOT'S ROOM
Maybe there is, Fur... I have an idea!
What if we use Root's shrink ray to go inside her and find out what's making her sick?
Whoa, weren't you supposed to give that back?
I was going to put it back in the lab... but instead I found these cool communicators!
Fin, do we need to have a talk about "borrowing" Root's things?

Moments later...
I'm gonna jump. Zap me when I'm over Root...
Z
Z
Z
Z
GET WELL SOON!
Okay...
ZAP!
GET WELL SOON!

Look out below!
Alright, Fur, I'm inside Root's body!
Great! What do you see?

VIRUSES
I see these things that look like little spiders. They're tiny!
I think we may be dealing with a **virus!**
In Root's basement laboratory...
VIRION
A virus is a microscopic **organism** that must infect a living cell to reproduce.
That little thing is called a **virion...**
Virion? What's that?

Virions are the particles of a virus.
They hold the virus's **genetic material,** which they inject into a host.
There goes one now, Fur! It's putting something into a cell!
It's injecting its genetic material into the cell, Fin!
The genetic material takes over the cell's machinery...
...using it to make more virions.
So, instead of performing its normal life functions, the cell manufactures more virus particles.
Ew, gross!

This says viruses can infect all forms of life...
...including animals, plants, and **microbes**–other microscopic living things...
Infection with a virus is the cause of many illnesses...
...from the common cold to more serious diseases, such as influenza and smallpox.
ACHOO
PARTS OF A VIRUS
Genetic Material
Capsid
Coat
Other viral infections can be harmless. What's the virus doing?

Can viruses reproduce on their own?
See page 40 for the answer.

BACTERIA
BACTERIUM
Bacteria are simple organisms that consist of one cell.
I think you're actually looking at **bacteria!**
There are bacteria everywhere... What should I do?
Now we know what's making Root sick...
Maybe we can figure out how to make her better.

Okay, Fin! Let's get you out before Root wakes up!
Got it!
Z Z Z
Soon...
SPLOOSH!
Oh...hello. You're still here?
We, uh, have a strong hunch that you have a bacterial infection.
You used my shrink ray to look inside me, didn't you?

Back in the laboratory...
What did the bacteria look like?
Some of them looked just like this one...

COMMON BACTERIA FOUND IN HUMANS:
Firmicutes (gut, mouth)
Staphylococcus (skin)
Bacteroidetes (gut, mouth)
Fusobacteria (mouth)
Proteobacteria (skin, mouth)
That looks pretty harmless...
Harmless bacteria are found in all healthy plants and animals...

In a healthy organism, these bacteria help crowd out harmful bacteria...
Did you notice anything else?
There were also a bunch that looked like this...
Ah, these are the bad guys! This should be easy to treat!
How can all of those things be bacteria? They look so different!

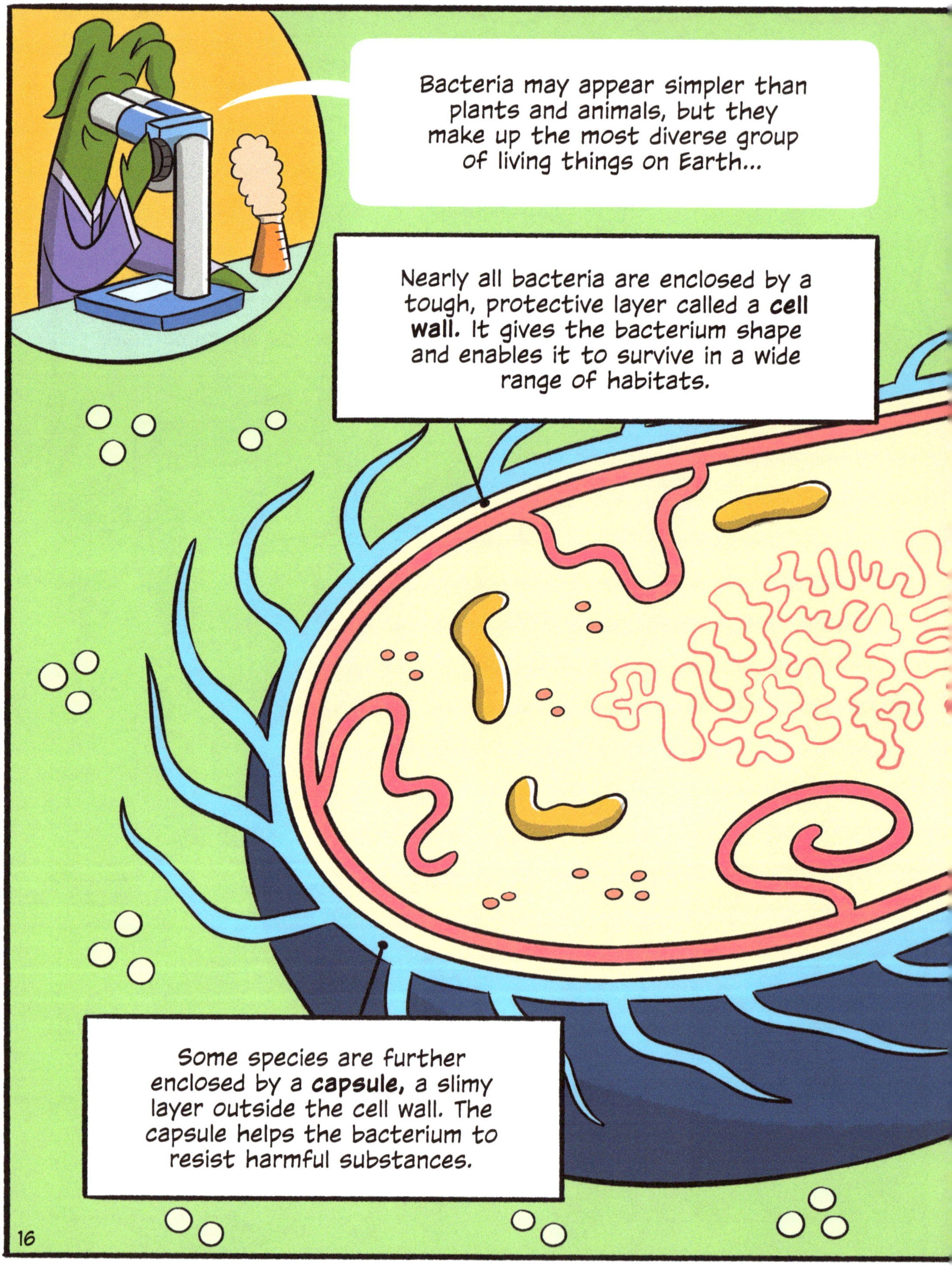

Bacteria may appear simpler than plants and animals, but they make up the most diverse group of living things on Earth...
Nearly all bacteria are enclosed by a tough, protective layer called a **cell wall.** It gives the bacterium shape and enables it to survive in a wide range of habitats.
Some species are further enclosed by a **capsule,** a slimy layer outside the cell wall. The capsule helps the bacterium to resist harmful substances.

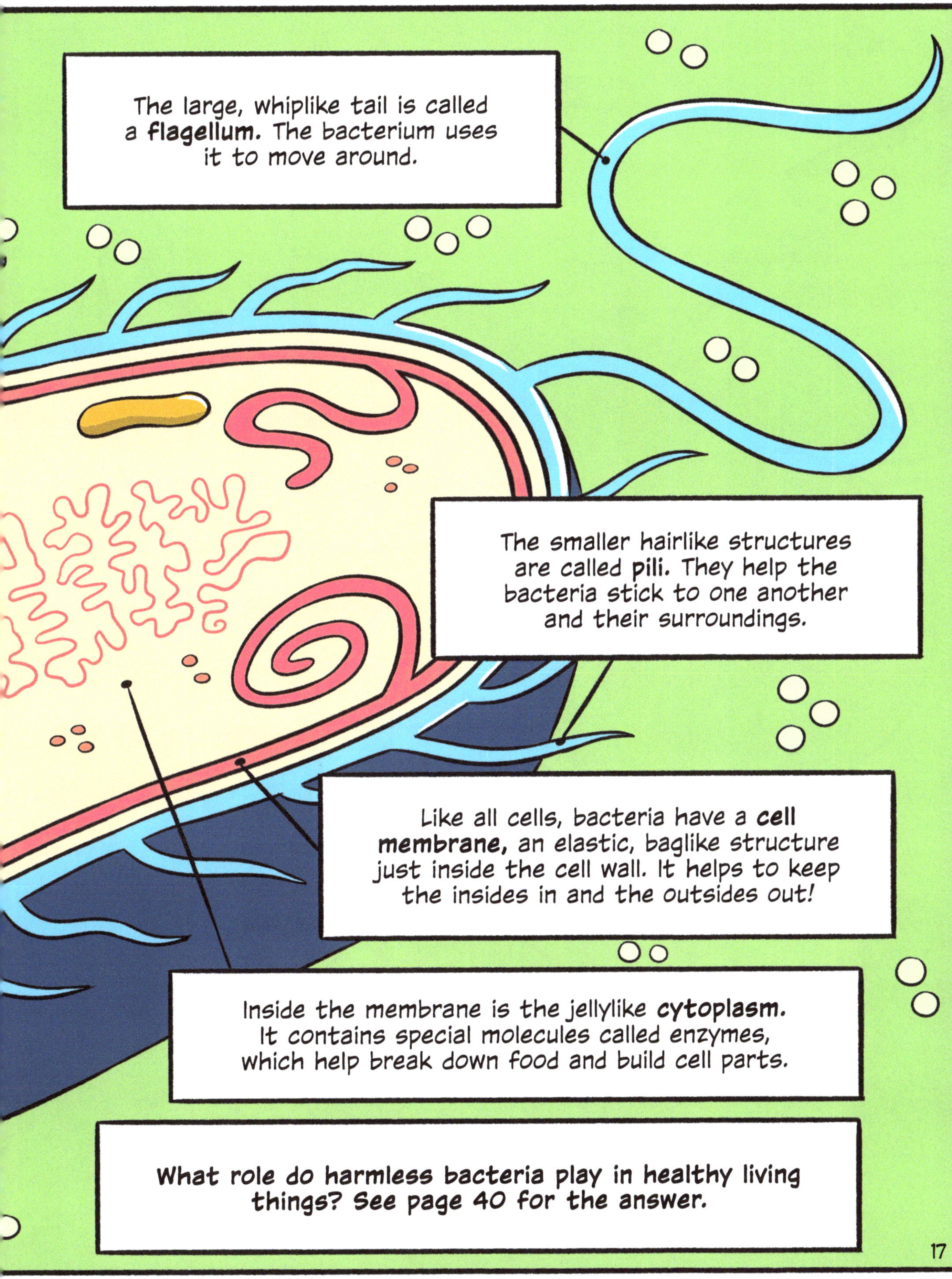
The large, whiplike tail is called a **flagellum.** The bacterium uses it to move around.
The smaller hairlike structures are called **pili.** They help the bacteria stick to one another and their surroundings.
Like all cells, bacteria have a **cell membrane,** an elastic, baglike structure just inside the cell wall. It helps to keep the insides in and the outsides out!
Inside the membrane is the jellylike **cytoplasm.** It contains special molecules called enzymes, which help break down food and build cell parts.
What role do harmless bacteria play in healthy living things? See page 40 for the answer.

Fortunately, my bacterial infection is easily treated with **antibiotics.**
Antibi-what now?
I'll call my friend Spore. He'll explain.
Later that day...
Greetings!
Hello, Spore!

Sorry to hear you're feeling sick, Root!
It's just a little bacterial infection... easy to treat–with a little help from **fungus!**
Fungus can help you get better, Root?
That's right!

Fungi, like our mushroom friend here, obtain food by absorbing it from other living organisms or parts of once-living things.
HUMMM
WHIRRR
CLICK!
We are neither plants nor animals, having diverged from them over 1 billion years ago.
Some of us consist of only a single cell. Take these molds, for example.
They produce molecules that destroy many kinds of bacteria, but are harmless to larger living things...
These substances are called antibiotics.
WHIRRRRR
HUMMMMMM

Here you are, Root.
Take some every day for the rest of the week, even if you feel better earlier.
Warning: Spore is an expert. Don't just drink molds you find growing around the house!
CLICK
I'm feeling better already!
Wonderful! Let's go for a walk. The fresh air might do you good.

In addition to single-celled fungi, many of us have multiple cells...
The most familiar are probably us mushrooms.
Mushrooms are actually the "fruiting bodies" of many fungi... kind of like flowers on a plant.
Only instead of making seeds, we release **spores** from gills or tubes on our undersides...

The rest of our life functions are carried out by a hidden network of threadlike cells that you can't see, sometimes called **mycelium.**
These cells *decompose* (break down) organic material and absorb its nutrients.
Fungi are incredibly important to the environment.
They decompose dead animal and plant matter, returning nutrients to the soil.
Oh, stop!
Mushrooms can also be eaten as food...
What do fungal cells look like, Spore?
Excellent question! Let's head back to the lab to study one.
True or false? Mushrooms are the parts of fungi that decompose organic material. See page 40 for the answer.

Back in Root's laboratory...
Fungal cells resemble plant and animal cells in many ways...
After all, we're all *eukaryotes*, living things whose cells contain a nucleus.
Nucleus
Mitochondrion (plural, mitochondria)
Bud scar
Cytoplasm

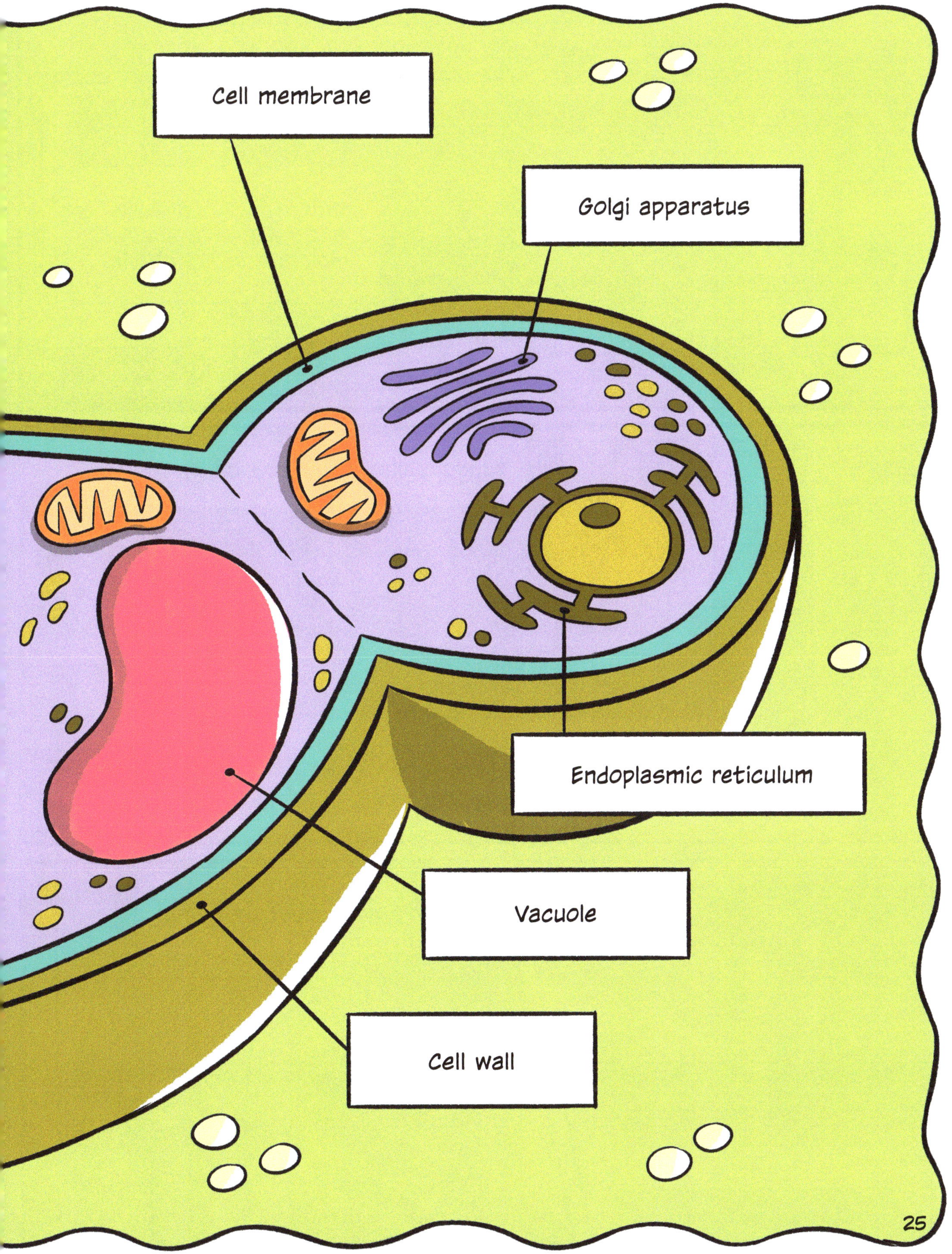
Cell membrane
Golgi apparatus
Endoplasmic reticulum
Vacuole
Cell wall

Fungal cells have cell walls, much like plant cells...
Nucleus
Cell wall
But, fungal cell walls are made of a special molecule called **chitin.**
Fungal cells can reproduce by budding...
Nucleus
Cell wall
Bud scar
The process leaves a chitinous mark, called a **bud scar,** on the surface of the parent cell.
Oh, look at the time! I must be going!
BOING!

Pleasure to meet you, Fin and Fur! Do feel better, Root! Farewell!
Wow! What a fun guy!
Yes! But not all fungi are so nice, you know.
Fill in the blanks. Fungi can reproduce through a process called _____ . This can leave a chitinous mark on the parent cell called a _____ . See page 40 for answers.

Some fungi can cause waste, disease, and even death.
GASP
Fur, you fluffhead! What did you do?!?

That rose was infected with a super-fungus I've been working on!
The fungus grows quickly on a wide variety of hosts.
Fortunately, I also invented a **fungicide** to control it. Suit up!

We've got to find that rose!
Follow the trail of fungus!
Look—the fungus has infected these insects.

Fungus can also infect plants. I engineered my fungus using genes from ergot.
Ergot is a fungus that infects wheat crops—and poisons bread made with them.
Why did you create this fungus, Root?
I guess I was just curious... I didn't really think it through.

Look! The super-fungus must have ruined this picnic lunch!
There's the rose!
Give me that!
Hey!

FOOSH!
Threat contained! We should continue to disinfect the park for a while though.
No problem... This is fun!
The end.

LIFE on the EDGE
Ideas from the Cutting Edge of Biology
MYCELIUM COMPOSITES
Remember that threadlike network of fungal cells called mycelium?
It can bind to the substance the fungus grows in, forming a material called a **mycelium composite...**
The composite's properties—such as how strong or heavy it is—depend on the type of fungi, the **growth medium,** and growing conditions...
WHIRRRRRR

Designers have made bricks, furniture, tiles, and other products out of mycelium composites.
Such products don't harm the environment and can even be composted.
Architects have imagined using mycelium composite as a building material.
This idea is called *mycotecture*—combining *myco*, meaning *fungi*, and *architecture*!
Mycotecture has even been proposed as a way of "growing" habitats for astronauts on the moon and Mars!

SCIENCE FUN WITH FUR!

THE GREAT MOLD RACE

YOU WILL NEED:

- eight small pieces of fresh bread (two large pieces cut into fourths should work)
- water and brush or spray bottle
- plastic wrap
- two paper plates

1
Spray or lightly brush four pieces of bread with water.
2
Wrap two of the moistened pieces and two of the dry pieces in plastic wrap.
WET WRAPPED
DRY WRAPPED
WET UNWRAPPED
DRY UNWRAPPED
3
Now that you've prepared your samples, place one of each kind on each plate. (Be sure to label your samples!)
4
Now set one plate on the counter and the other in the refrigerator.
Check the bread each day, looking for mold. In which conditions did mold grow fastest?
Gross, Fur! What are you doing?

SHOW WHAT YOU KNOW

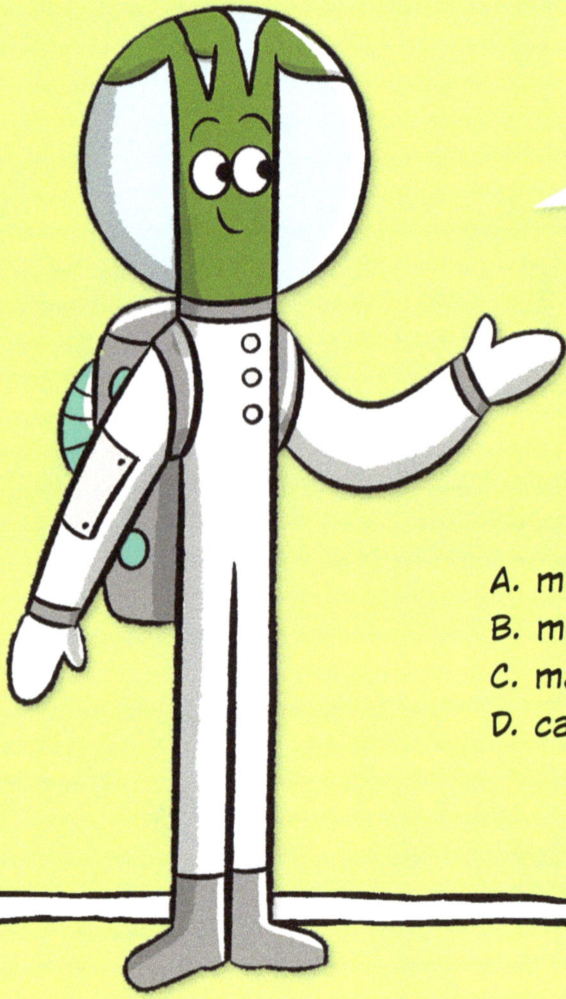

1. Does each of the following describe bacteria, fungi, or viruses?

A. must infect a living cell to reproduce
B. most diverse group of living things
C. made up of particles called virions
D. can consist of one or many cells

2. Fill in the blanks.

Some fungal cells produce special medicines called _____ . These can be helpful in killing _____ .

3. What is the role of a mushroom in fungi? What is the role of the mycelium?

4. Name these parts of the bacterial cell.
A. tough protective covering around the cell
B. tiny hairs that help the bacterium stick to its surroundings
C. whiplike tail used for movement
D. slimy covering that protects the bacterium against harmful substances
See page 40 for answers.
39

ANSWERS

page 11: No, a virus cannot survive on its own. It must infect a living cell to reproduce.

page 17: In healthy living things, harmless bacteria help to crowd out harmful bacteria.

page 23: False. The mycelium decomposes organic materials. Mushrooms are the fruiting bodies that spread spores, helping fungi to reproduce.

page 27: budding; bud scar

SHOW WHAT YOU KNOW ANSWERS pages 38-39:

1. A. viruses
   B. bacteria
   C. viruses
   D. fungi
2. Antibiotics; bacteria
3. Mushrooms are the fruiting bodies that spread spores, helping fungi to reproduce. Mycelium breaks down organic material for food.
4. A. cell wall
   B. pili
   C. flagellum
   D. capsule

WORDS TO KNOW

antibiotics a medicine made by fungi and used to fight bacteria.

bacteria (singular, bacterium) a simple living thing consisting of one cell.

bud scar in fungi, the chitinous mark on the surface of a parent cell left behind by budding.

capsule a slimy outer coating that protects bacteria from harmful substances.

cell membrane the thin, flexible covering of a living cell.

cell wall the hard outer covering of cells in certain types of living things.

chitin a tough substance found in the cell walls of fungi.

cytoplasm the jellylike fluid that fills a living cell.

flagellum (plural, flagella) a tiny, whiplike structure that certain cells use to move.

fungicide a substance that kills fungi.

fungus (plural, fungi) living things that obtain food by absorbing it from other living things or from parts of formerly living things.

genetic material structures that carry the genetic "instructions" for making a living thing.

growth medium a substance in which living cells are grown.

microbe a single-celled organism.

mycelium in fungi, a tangled mass of threadlike cells.

mycelium composite a material consisting of mycelium and a growth medium.

organism a living thing.

pili (singular, pilus) in bacteria, a hairlike structure that helps one bacterium attach to another.

spore a reproductive cell.

virion a virus particle.

virus a tiny living thing that must infect another living thing to reproduce.

www.ingramcontent.com/pod-product-compliance
Lightning Source LLC
LaVergne TN
LVHW070859210926
848482LV00016B/974
9780716667537